Michael Zhilaev

N.SEMENOV

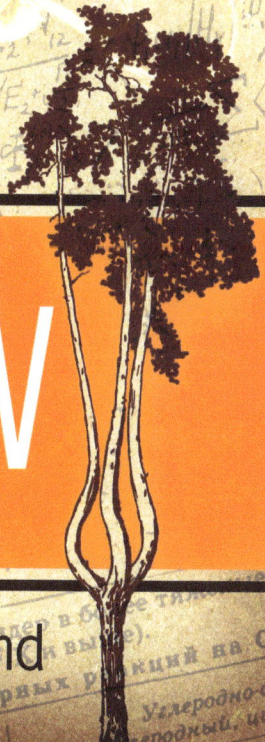

Soviet Science Legend

To my loving sister Olga, with gratitude.

Ignore the past – you lose an eye

Forget the past – you lose two eyes

— Old Russian Proverb

CONTENTS

INTRODUCTION

Approximately a one-hour drive north of Moscow, in the middle of a pine forest, stands a sign: Chernogolovka–Noginsky Scientific Center. Across from a group of tall modern buildings next to the bus station is the House of Scientists which serves as a movie theater, concert hall, and gathering place for locals. The monument to the founders of Chernogolovka is located right here. Two men stand together, ready for the start of something big. Semenov is explaining his plans while his associate Dubovitsky has already rolled up his sleeves to start working. This monument was created in 2016 to commemorate Semenov's 120th birthday and the 60th anniversary of him receiving the Nobel Prize.

The story you are about to read is about Nikolai N. Semenov—the man behind the creation of the Noginsky Science Center in Chernogolovka and the Soviet Atomic Project.

Figure 1. Monument to Noginsky Scientific Center's founders:
Semenov (left) and Dubovitsky (right).

RESTLESS EXPLORER

YOUTH ON THE VOLGA

Nikolay Semenov's parents, Elena and Nikolay, moved from Tsarskoe Selo (present day Pushkin), a town on the outskirts of St. Petersburg, to the large city of Saratov in 1895 shortly before the birth of their son. His father received a title of nobility from the tsar while his mother Elena, who was educated at Bestuzhev Gymnasium in Petersburg, was a math teacher. Shortly after his tenth birthday, young Kolya—as Nikolay was called by those close to him—started attending a local school in Volsk, a town near Saratov on the banks of the Volga River. As a fifth-grade student, Semenov fell in love with chemistry. His family would order books directly from St. Petersburg because Kolya was very eager to learn more than was taught at school.

Figure 2. Local school (now school #16) in Volsk attended by Semenov from 1903 to 1910.

At that time, explosions were often heard from the field near the Semenov's house as the young explorer was attempting his first chemical experiments. One of these was to prove that the white powder that came from the reaction between chlorine and sodium was table salt. After he completed the experiment he took the powder, spread it over a slice of bread, and ate it.[1]

Two years later Nikolay's family moved to Samara, another large city on the Volga. There he became a student at the local school—one that emphasized the sciences over classical literature and philosophy. Here Semenov met someone who would go on to greatly influence the future of his career: Nikolay Karmilov, a young teacher passionate about physics who had just graduated from Kazan University. After seeing the limitations of the school program, Karmilov started a home university at Semenov's house with full support from Nikolay's mother. Students conducted experiments in physics and chemistry beyond the college program. Semenov's mother recalled frequently hearing explosions from the students' lab. Karmilov's devotion to science was contagious and instilled a curiosity and love of science into his students.[2]

Even when Semenov became a student majoring in math and physics at St. Petersburg University in 1913, his lifelong friendship with Karmilov continued despite the physical distance between them. After the war, Karmilov's daughter, and later her children, worked at the Institute of Chemical Physics (ICP) in Moscow.[3]

AFTER COLLEGE: CIVIL WAR EXPERIENCE

In 1917, Nikolay Semenov graduated from St. Petersburg University with a degree in physics. He was later invited to continue working at the university, however the start of the Bolshevik revolution shook the entire country. Through the winter of 1918, men and women were dying from hunger in Petrograd (St. Petersburg was renamed Petrograd in 1914 at the beginning of World War I to eliminate German words in the city name). Nikolay quickly left the city to visit his parents in Samara. Unfortunately the Russian Civil War broke out in the summer of 1918. The whole country was divided: revolutionaries fought for the newly created Red Army while those who supported the Tsarist government fought in the White Army. Nikolay, focusing entirely on science, was not interested in the political situation. He was drafted into the White Army in Samara but soon deserted it and moved to Tomsk, Siberia, where he continued his research at Tomsk University with Professor Weinberg. The following year he was sent back into the White Army again, but with the help from Tomsk University soon returned back to his studies.

Semenov's desertion and role in the White Army were well known to Soviet officials which endangered his life, especially during Stalin's repression in 1937. During this time, many scientists and their family members, such as theoretical physicist Landau, were arrested by the government, while others such as Semen Shubin and Matvei Bronstein were killed.[4]

SCIENCE: FIRST CONNECTIONS

Despite tremendous economic problems after World War I, the October Revolution, and the Civil War, the advancement of science was one of the top priorities of the new Communist government. Soon after the war ended, Semenov was recalled back to St. Petersburg by his former teacher and mentor Professor Ioffe. Ioffe's scientific seminars were a gathering of young and talented scientists. Many students called him "Papa Ioffe" and regarded him as their teacher. These seminars and workshops played an essential role in disseminating and discussing new discoveries in physics. Semenov knew Ioffe from his second year at the university and had done some experiments with him. While still a student, he published his first scientific work alongside Ioffe.

Semenov recalls that in 1921, shortly after WWI and the Civil War were over, Ioffe purchased advanced scientific physical and chemical equipment for his Physical-Technical Institute with the support of the government.[5]

During one of Ioffe's seminars, Semenov met another young scientist, Pyotr Kapitsa. They soon became close friends and

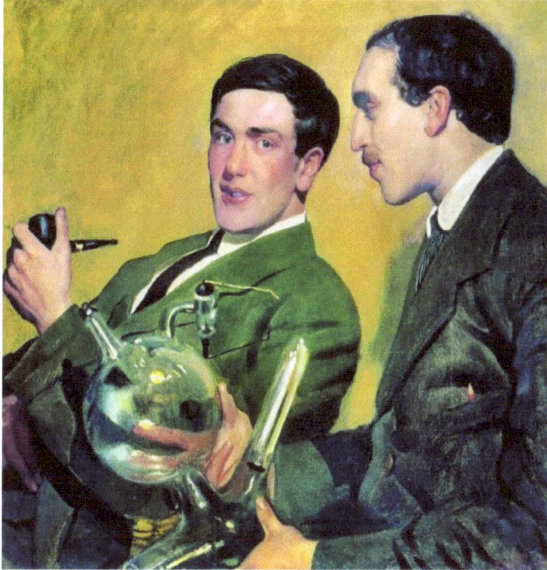

Figure 3. Kustodiev: Portrait of Pyotr Kapitsa and Nikolay Semyonov. 1921.

started working together under the guidance of Ioffe. Soon, they published their first research paper together, "On the Possibility of Experimental Determination of Magnetic Movement of an Atom". Their friendship is even commemorated in two famous paintings by Kustodiev, one of the most outstanding portraitists of Russia.

Many Russians are familiar with the "*Portrait of Pyotr Kapitsa and Nikolay Semyonov*" by Kustodiev. What they do not know is the story behind this painting. In 1921, Kustodiev was working on a portrait of a well-known opera singer, Feodor Chaliapin, when two young scientists simply came by and said, "*You are known for painting celebrities. Try to paint our portrait because we will become famous.*" An astonished Kustodiev stopped all his unfinished work and painted a portrait of Semenov with Kapitsa. "*And they are red-faced, and so cocky and funny that I had to agree,*" remembered Kustodiev. "*They brought the x-ray tube, which worked in his Institute, and it went. Then the fee is brought in, you know what? A rooster and a bag of millet. Just earned these somewhere near St. Petersburg, after making some proprietor of the mill.*" [i]

---

[i] https://arthive.com/boriskustodiev/works/379818~Portrait_of_Pyotr_Kapitsa_and_Nikolay_Semyonov

5

Figure 4. Kustodiev "Festivities in Honor of the Second Comintern Congress on 19 July 1920", 1921.

The young men did not deceive Kustodiev. Years later both became Nobel laureates. Pyotr Kapitsa (on the left with a smoking pipe) earned the honor in physics. The other, Nikolai Semenov, was awarded the prize in chemistry. Science does not know the answer to the question of what was responsible for their wins. Were the young scientists so confident that Kustodiev decided to seek a bit of glory in advance? Or did Kustodiev's brush become a blessing to them and brought them good luck? Interestingly, the two young friends appeared in another Kustodiev painting - *"Festivities in Honor of the Second Comintern Congress on 19 July 1920. Demonstration on Uritsky Square (now Palace Square)"*.

The friendship between Kapitsa and Semenov lasted throughout their lifetimes, in good and challenging times. In 1936, when Kapitsa returned to the Soviet Union from the United Kingdom and was prohibited from leaving the country again, Semenov supported his friend.[6] Later, during a period of isolation in Kapitsa's life, Semenov's wife, Natalia, visited the exiled friend in his suburban home near Moscow. Risking her life and her family's lives, she saved Kapitsa's archive under her bed for eight years. This archive contained very important correspondences from Kapitsa with Stalin, Molotov, Beria, and other Soviet leaders.[7]

6

In the picture below from the archives of Semenov's daughter,[8] Semenov is visiting Kapitsa at his suburban house at Nikolina Gora, also known as a "dacha."

Connections with Ioffe and Kapitsa were fundamental to Semenov's scientific establishment. In 1921, when Nikolay Semenov had just started working with Ioffe, he met a student of the Petrograd Polytechnic Institute (now Peter the Great St. Petersburg Polytechnic University), Yulii Khariton, and invited him to his research group. This was the start of his first laboratory which consisted only of three physicists: Kondratiev, then nineteen, Khariton, seventeen, and Walter, twenty-one.[9] Years later, Yulii Khariton would be named the father of the Soviet atom bomb.[10]

BRANCHING CHAIN REACTIONS

In 1925, Semenov's laboratory staff, Z. Walta and Yu. Khariton, made their first major discovery while exploring the conditions in which white phosphorous, a substance that glows at room temperature, would stop glowing.[11] These experiments

Figure 5. Kapitsa and Semenov are resting at Kapitsa summer house of Nikolina Gora, a "dacha".

7

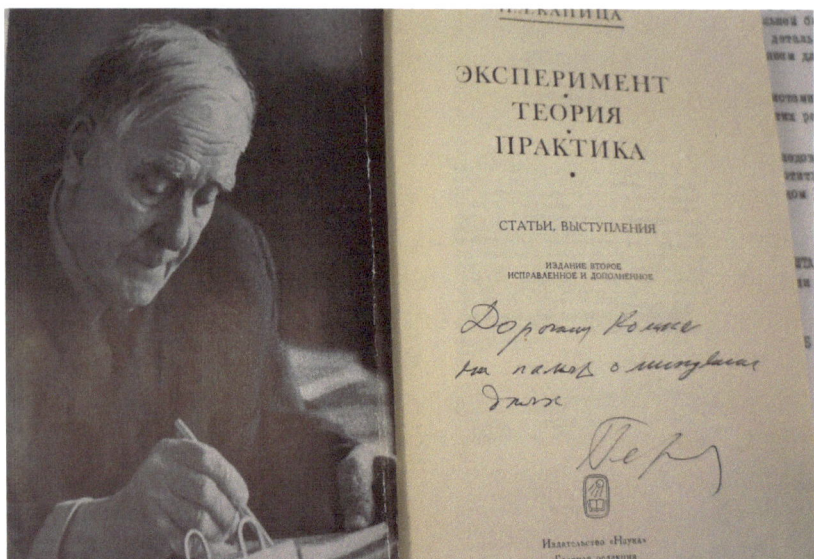

Figure 6. The picture reflects the close friendship between them; Kapitsa is using a nickname "Kolka" in his book signed to a lifelong friend. The book belongs to the Semenov Museum in Chernogolovka.

led Semenov to formulate the theory of branched chain reactions. While Oxford scientists, Hinshelwood and Thompson, had discovered the upper pressure limit in the explosive chain reaction,[12] Semenov and his laboratory were focusing on investigating the relationship of pressure to this process and discovered low pressure explosive limits.[13]

Both Hinshelwood and Semenov came to the same explanation for why the same reaction can proceed either explosively or slowly based on branching chain reactions. The scientists received the Nobel Prize in 1956 for their combined work.

The theory of chain reactions had huge practical implications for research in combustion, oil cracking, and many other processes. The fact that even 5 percent methane leakage can lead to an explosion caused by either a spark or even turning off the light can be explained by Semenov's Theory.[14]

Semenov continued his experimental and theoretical work on branched chain reactions. In 1928, he determined parameters for

The Nobel Prize in Chemistry 1956

Sir Cyril Norman
Hinshelwood
Prize share: 1/2

Nikolay Nikolaevich
Semenov
Prize share: 1/2

Figure 7. Nobel Prize winners in Chemistry in 1956: Hinshelwood and Semenov.

thermal explosion and flame propagation caused by any exothermic reaction. He wrote down his theoretical and experimental research in a paper, "Thermal Theory of Combustion and Explosion," published before WWII.[14,15]

ATOMIC BOMB RESEARCH

In 1927, the still young Semenov was appointed to lead a department at the Leningrad Physical-Technical Institute (now The Ioffe Physical-Technical Institute of the Russian Academy of Sciences). In 1931, his department became a separate entity: The Institute of Chemical Physics. Semenov remained the director of this institute for over fifty-five years until his death in 1986.

From 1926–1928, Semenov's colleague and pupil Yulii Khariton, on Pyotr Kapitsa's recommendation, became a member of the Rutherford Laboratory at Cambridge University. On his way back to the Soviet Union, he made a stop in Berlin to visit his mother. Khariton was shocked by the Nazi Germany's military strength and propaganda calling for war. Back home in Leningrad, he also noticed that many young people were wearing military uniforms, marching, and training in military activities such as shooting and parachute jumping. He felt that science needed to react to this situation and decided to research new and more powerful explosives to help his country prepare for the inevitable conflict.

By 1939, Khariton, while working with Zeldovich, discovered that the reaction of splitting the uranium atom agreed with Semenov's branching chain reaction model. Together, these two scientists calculated the chain reaction of fission in uranium. The results of the Khariton-Zeldovich experiments all pointed to a revolutionary discovery in nuclear physics. However, Semenov's report about these discoveries to Soviet authorities was not immediately answered. Only in 1942, after Soviet spies obtained the information on full-speed nuclear research development in England and the United States, Semenov was contacted by the Soviet Defense Committee to work on a dedicated Government Nuclear Committee.[16]

Work on the Soviet atomic bomb began soon after, with many theoretical preparations already complete before the start of WWII. However, progress on nuclear research was temporarily disturbed by the war, during which Semenov's institute worked for the immediate needs of the Soviet defense industry. St. Petersburg (renamed Leningrad) was under German siege for almost nine hundred days, so in late August of 1941, the Institute of Chemical Physics was evacuated to Kazan and later, in 1943, relocated to Moscow.

THE SOVIET ATOMIC PROJECT

Semenov's group resumed its work on the atomic project in 1943. Although America and Britain were allies with the Soviet Union in the war against the Axis Powers, the Soviet Union remained a communist regime and a potential future enemy.

During the 1943 Quebec meeting, Churchill and Roosevelt agreed to join their efforts in a single "Manhattan Project" to accelerate the preparation for an atomic bomb, and keep it hidden from the Soviet Union. However, by that time Soviet spies had already infiltrated the Manhattan Project. The most critical design information of the first American plutonium-based bomb, "Gadget," was obtained only twelve days after the bomb's assembly. In his interview with V. Sojfer, Y. Khariton admitted that the progress of the Soviet Atomic Project was greatly accelerated with the help of communist scientists from the American/British Atomic Project: British physicist Klaus Fuchs (code name Mlad) and American physicist Ted Hall (code name Perseus). Khariton said that the first Soviet nuclear bomb had been a copy of the first American atomic bomb.[17]

Fast forward to 1959, when Fuchs was released after serving nine years in British prison, Khariton had contacted the Soviet government with a petition to award Fuchs for his contribution to the Soviet Atomic project, but his request was denied.[18]

Figure 8. Klaus Fuchs (1911-1988), a German theoretical physicist and spy who worked at Los Alamos during the Manhattan Project and passed atomic secrets to the Soviet Union.

After the United States dropped an atomic bomb on Hiroshima in August 1945, Stalin turned the existing Soviet atomic project into a frantic effort as he gave very little time to construct an atomic bomb twice as powerful as the "Fat Man" and "Little Boy" bombs dropped on Hiroshima and Nagasaki. It was August 20, 1945, just eleven days after the bombing of Japan, when Commissar General of State Security L. Beria organized and led a special Nuclear Committee to create an atomic weapon for Soviet Union. The first two years of theoretical work on the Soviet Atomic Project (1946–1947) were conducted in a Special Sector of the ICP in Moscow led by M. Sadovskii. The sector work on the methodology, scientific equipment, and training resources were as essential to prepare for nuclear experiments. The special sector work was classified: this group did not publish results of their nuclear explosion studies and did not take part in any international scientific communications.[19]

FROM THEORY TO TESTING SITES

Parallel with theoretical and computational work, Soviet authorities created an experimental scientific team called Lab #2, which would later become the Institute of Atomic Energy. This nuclear scientific research group was led by Igor Kurchatov, with technical leadership assigned to Y. Khariton.

In April 1946, the leaders of Lab #2 decided to create another group, a top secret nuclear laboratory (work name KB - 11) located near the city of Sarov in the deep woods of the Nizhny Novgorod region which became the first Soviet Nuclear Center, the Soviet equivalent of American Los Alamos. In February of 1948, Semenov provided the government a list of ICP scientists to be relocated to Sarov for one year of work on resolving the Soviet Atomic problem. Of the fifteen scientists working in KB-11, seven outstanding researchers were from Semenov's Institute: Y. Khariton, Y. Zeldovich, K. Shelkin, V. Bobolev, A. Belyaev, D. Frank-Kamenetskii, A. Apin, and G. Gandelman. An atmosphere of extreme confidentiality surrounded Sarov. Starting 1947 the city was excluded from all geographical maps. Center -300, Arzamas

Figure 9. Sarov KB-11 – Yu. Khariton and his first Soviet Union Atomic Bomb RDS-1 states for Russian self-made, "Rossiya Delaet Sama". Picture from https://www.atomic-energy.ru/.

-75, Privolzhsk, and Kremlev were names used to refer to Sarov or Arzamas-16.[20]

The physicists working there could only dream about visiting their homes, going on family vacations or even attending their parents' funerals.

At the same time, during the summer of 1946, Semenov initiated the construction of the second testing site. In his letter to Beria, Semenov demonstrated all available metrics to build the case for a new experimental base. The suggested sites were three areas in Kazakhstan. After carefully verifying the Semenov proposal in April 1947, the Special Committee officially approved a project (Code name – Gornaya Stantsiya) to create a second experimental base just 140 km west of Semipalatinsk, Kazakhstan. The location of the Semipalatinsk experimental base, also known as "The Polygon" was chosen based on the nearby airport, transportation along the Irtysh River, and atomic industry facilities in the Southern Ural Mountains. At the same time, the base was at a remote distance from populated areas. On June 19, 1947, the Soviet government issued the "strictly confidential" decree #2142-564 for ICP director

Figure 10. First Soviet Atomic Bomb Explosion. Semipalatinsk, August 29, 1949.

Semenov, the director of Gornaya Stantsiya, Sadovskii, and the First Department (Vannikov, Zavenyagin, Alexandrov). The decree stated that the organization would need to provide the technical specifications for all technical equipment for airplanes to the Soviet Ministry of Defense within twenty days. That marked the start of construction of the nuclear base Semipalatinsk -21 which remained in operation for forty years with 456 atomic and hydrogen explosive experiments performed there.[21]

The building of the atomic bomb and the Semipalatinsk test ground was conducted quickly and under pressure for time. Within 14 months, the Semipalatinsk testing ground was ready for the first atomic test with 48 types of special devices built by ICP scientists to measure the nuclear explosion.[ii] On Aug 26, 1949, Commissar General of State Security L. Beria, who personally supervised the Soviet Atomic project, submitted the first Soviet atomic bomb execution plan developed by Kurchatov, Vannikov and Pervuhin for approval to Stalin. However, that plan did not have Semenov's name on it (see Appendix 1).

[ii] Г. Киселев. (2015). Научное наследие лауреата Нобелевской премии Н.Н Семенова в Советском Атомном проекте. Документы и воспоминания. Sarov.

Figure 11. Khariton (left) with Semenov – 1980s.
Photo credit – A. Semenov.

Unfortunately after WWII, Semenov became the victim of personal attacks from Akulov and other "patriotic" scientists, mainly due to his close relationship with Pyotr Kapitsa who was known for his radical political opinions and international connections.[iii] At that time, Semenov's loyalty to the Soviet government was under suspicion. These circumstances decreased his influence on the Soviet Atomic Project. It was humiliating for him to be restricted from participating in the first atomic bomb RDS-1 test in Semipalatinsk, Kazakhstan, on August 29, 1949. Semenov's close colleagues who attended the Semipalatinsk nuclear test were prohibited from giving Semenov any classified information. Only after Stalin's death in 1953, and when Semenov became a Nobel Prize winner in 1956, was his influence on the scientific community fully restored.

Soon, after the successful detonation of the American-designed bomb in 1949, Khariton and his group restarted their original mission to create an even more effective explosive. They came up with the plutonium model, twice as powerful as the previous bomb.

[iii] Семенова Л.(1993).То Что Всегда со мной. Сборник Воспоминаний о Николае Николаевиче Семенове. Москва, с. 198

Figure 12. Semenov (right) with his pupils and collaborators: Y. Zeldovich and Khariton. Photo credit – A. Semenov.

NOVAYA ZEMLYA

After the first ground-based atomic/hydrogen bombs were successfully detonated in Semipalatinsk, scientists turned to the sea to conduct further nuclear tests. An archipelago in the Arctic Ocean located in northern Russia, Novaya Zemlya, was chosen as the second site to test large nuclear explosions after the Soviet Union determined that its test site in Kazakhstan was too close to human settlements. On July 1 of 1954 the Soviet government issued a decree to create a test site at Novaya Zemlya (code name Object 700). It took exactly a year to build the test site. The first underwater test was successfully conducted in Chernaya Bay, part of Novaya Zemlya, on September 21, 1955.

Semenov was appointed to lead the development of a nuclear experimental base at Novaya Zemlya. From 1955 to 1961 scientists

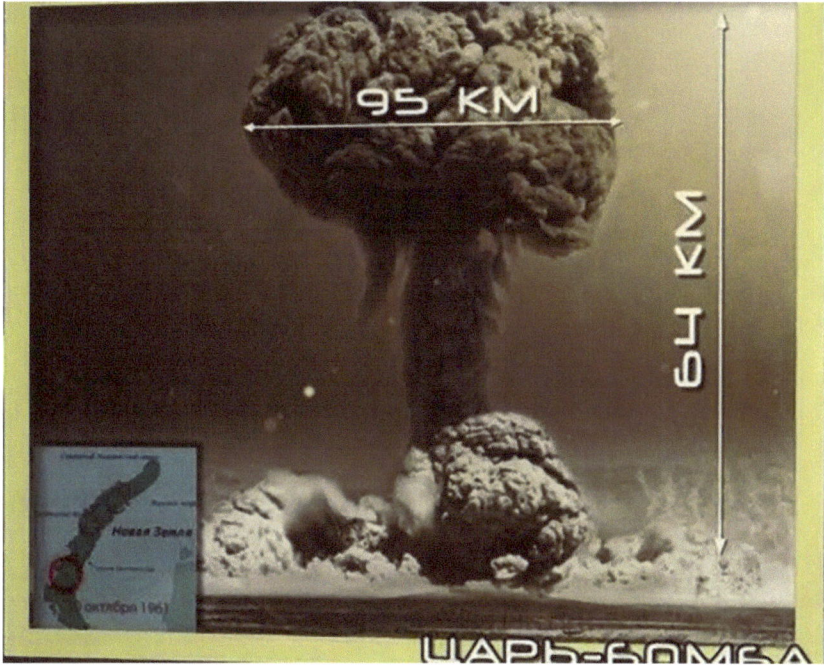

Figure 13. The explosion of "Tsar Bomb" (King of bombs) on Novaya Zemlya, October 30, 1961. Radiation parameters: cloud diameter – 95 km, could height – 64 km. Picture from www.historius.ru.

conducted several major tests at Novaya Zemlya under his supervision. In 1961, Semenov led a series of air-based nuclear tests at Novaya Zemlya.

AN-602, also known as Tsar Bomb, RDS – 220, Kuzka's Mother, and Big Ivan, was the most powerful thermonuclear explosive ever detonated. It was tested there on October 30, 1961.[22]

The next phase of development and testing lasted from 1961 to 1962 and consisted of high-altitude nuclear explosions. From 1957 to 1962, Semenov served as a chief scientific officer of programs to develop high-altitude nuclear explosions in order to research the qualifications and impact on air-based weapons, such as missiles and planes. The series of eight high-altitude nuclear explosions were conducted in the bare steppe Kapustin Yar, the Soviet rocket launch and development site in the Astrakhan region. ICP scientists Kompaneets and Nemchinov developed theoretical computations and the high-altitude equipment for these tests.

In summary, the ICP led by Semenov advanced the Soviet Atomic Project in the following ways:

1. Calculating the parameters of the atomic bomb during initial development.
2. Pioneering the research on utilization of nuclear energy for military and manufacturing.
3. Organizing experiments to investigate parameters of atomic and hydrogen bombs.
4. Design and engineering of testing bases at Semipalatinsk and Novaya Zemlya, and the equipment to measure various characteristics of nuclear and hydrogen explosions.
5. Conducting work fundamental to establishing protection from nuclear weaponry through methods using proton radiation.

Overall, the Semenov Institute developed and produced about 80 percent of all equipment for the nuclear tests from 1949 to 1963. This experimental research led ICP scientists to develop a theory of nuclear explosions in high altitudes as well as learn critical information about nuclear explosions in different environments.[iv]

[iv] Г. Киселев. (2015). Научное наследие лауреата Нобелевской премии Н.Н Семенова в Советском Атомном проекте. Документы и воспоминания. Sarov,c.16

CHERNOGOLOVKA RESEARCH CENTER

BEGINNINGS

The story of the Scientific Center in Chernogolovka begins with the construction of a scientific testing ground for the ICP of the USSR Academy of Sciences on the land belonging to the N. E. Zhukovsky Air Force Engineering Academy in the Noginsky District, about an hour (thirty-five miles) north of Moscow. In 1946, the Soviet government appointed the task of leading dedicated experimental and theoretical research on atomic weaponry to the ICP. It was very important to the government to quickly develop an atomic bomb capable of rivaling what the United States had done. Special equipment and new methods needed to be developed specifically for this research. In 1953, another decree was given to the ICP to begin new research and testing for the development of powerful explosive substances. In addition to the research done on the Atomic Project, the Institute was also ordered to begin studies on high-energy solid rocket fuel. All of these studies would necd to be conducted far from the research center of the ICP in Moscow due to the dangerous nature of the expcriments.[23] The hardest task was to choose the location for the testing ground for the high-caliber military experiments near Moscow. Semenov and his associate Dubovitsky had an opportunity to personally visit several airbases and picked the location next to the small village Chernogolovka.[24]

In 1955, Nikolay Semenov, a Nobel Laureate and director of the ICP, sent a request to the Deputy Chairman of the Council of Ministers, M. Pervuhin, detailing the need for the creation of a special testing ground for the Institute. As the new testing ground was to serve the military, the government enthusiastically approved this project, declaring that the construction should begin immediately and finish by December 1, 1958 (Appendix 2).

Figure 14. Fedor Dubovitsky.

The role of Deputy Director of Science at the Chernogolovka testing ground was given to Fedor Dubovitsky, who now had the responsibility of managing all activities on the base, both scientific and organizational.

In March 1956, the zone planning committee started choosing grounds for scientific experiments and housing with help of young ICP scientists. Semenov recruited four students in their mid-twenties: L. Stesik, G. Manelis, A. Dremin, and A. Merzhanov. These four became the first heads of new laboratories in Chernogolovka and, like Semenov and Dubovitsky, became Chernogolovka legends.[v] The construction work was conducted by the Army Builder unit led by General–Major N. Zolotarevsky.

In 1959, the testing grounds were ready to welcome their first occupants to the apartment building at #3 First Street, but

---

[v]   http://50-let.chg.ru/chronology.htm

Figure 15. Fedor Dubovitsky's car "Pobeda" parked next to the House #3 on the First Street, where he was living in Apartment #7. Picture from Chernogolovka.ru.

who would come to the middle of nowhere? Established ICP researchers were reluctant to move to Chernogolovka from the prestigious and comfortable Moscow. Forget about access to popular Moscow theaters! The Chernogolovka base was surrounded by a forest, and people were forced to deal with the lack of even the most basic needs, such as stable electricity, which came only from the diesel station. There was no regular transportation between the settlement and Moscow, and the only place to work was a wooden barrack shared with the construction soldiers. Unsurprisingly, the majority of the first tenants at the Chernogolovka base were chemists and physicists in their twenties who spent their childhood in the Great Patriotic War. Many of them started working as teens to provide for their families while some were evacuated to eastern parts of the country. Material deprivation, hunger, and relocation were common to most of these young scientists' upbringings.[vi]

[vi] Rogacheva, M. (2017). The Private World of Soviet Scientists from Stalin to Gorbachev. Cambridge University Press.

By the end of 1959, the Chernogolovka base had an apartment building, a cafeteria, a hotel for eighty people, and the armored casemate for explosives testing.

The purpose for the Chernogolovka testing ground was solely of a military nature to produce powerful explosives, but that was not what Semenov envisioned. He believed in multidisciplinary collaboration and imagined that it would happen best by building a science center similar to Cambridge or Princeton. After Stalin's death, Semenov cultivated a good relationship with the new Soviet leader Nikita Khrushchev.

From 1959 to 1962, Semenov and Dubovitsky brought many country and science officials to Chernogolovka. In 1962, when the newly elected President of the Academy of Science of the USSR Keldysh first visited the Chernogolovka branch, he was impressed by the number of fully functioning buildings dedicated to labs and mechanical shops and the comfortable and modern organization of the settlement. At that time, Chernogolovka Branch of ICP had eight hundred employees and had already produced its first revolutionary results in rocket fuel development. On August 9, 1962, M. Keldysh signed decree #141-1275 issued by the Presidium of Academy of Sciences to build a scientific center in Chernogolovka (Appendix 3&4). The ICP formed the basis of the Noginsk Scientific Center of the Academy of Sciences. The decree gave the Board of Directors the power to overlook all scientific and organizational matters in the newly established scientific center led by Semenov.[vii]

Chernogolovka became an example of the new generation of scientific towns that emerged under Khrushchev. Physics centers in Dubna and Troitsk, a biological center in Pushchino, and a center of microelectronics in Zelenograd all demonstrated the efforts of the Soviet government to produce scientific breakthroughs as a part of building Communism.

Dubovitsky's energy, creativity, personal connections, and involvement in every aspect of the young center areas generated

---

[vii] Дубовицкий, Ф. (1992). Институт химической физики: очерки истории. Черноголовка.

enthusiasm among the young scientists and contributed to the rapid development of Chernogolovka. For example, from 1959 to 1964, the testing ground did not have a grocery store. Dubovitsky was aware of this issue and tried different approaches to help scientists with food supplies. A small grocery store was created right at one of the institute's buildings. Many scientists remembered how, in 1963, the bus once brought a large barrel of red caviar to the institute. The barrel was put at the center of the grocery kiosk, and everyone could take as much as they wanted for a very low price. For many years, a parcel system of food supplies helped scientists with managing provisions. Once a week, ICP volunteers would bring in high-quality foods, such as meat, cheese, kielbasa, etc., from Moscow and distribute these among employees.[viii]

The following leading institutes of the Russian Academy of Sciences were established as part of the Noginsk Science Center in Chernogolovka[ix]:

1961 – The Branch of the Institute of Chemical Physics, renamed in 1997 to the Institute of Problems of Chemical Physics

1963 – The Institute of Solid State Physics

1964 – The Institute of New Chemical Problems

1965 – The Landau Institute of Theoretical Physics

1969 – The Institute of Experimental Mineralogy

1972 – The Experimental Factory of Scientific Supplies

1978 – The Institute of Physiologically Active Substances

1983 – The Institute of Problems of Microelectronic Technologies

1986 – The Institute of Energy Problems of Chemical Physics

1987 – The Institute of Structural Macrokinetics

[viii] Rogacheva, M. (2017). The Private World of Soviet Scientists from Stalin to Gorbachev. Cambridge University Press.

[ix] Дубовицкий, Ф. (1992). Институт химической физики: очерки истории. Черноголовка.

TEACHER AND ENABLER

Semenov believed that what makes a researcher a true scientist is more than just talent. "Unlike in music and art, where talent is critical and usually discovered very early, it takes good teachers and family education to develop and nurture a talent in research, and it is usually discovered much later than artistic talent." According to Semenov, in science, dedication, stamina, perseverance, and the ability to analyze failures are even more critical than talent.[x] Throughout his life, Semenov discovered and helped many talented young scientists who regarded Semenov as their teacher: Y. Khariton, V. Kondratiev, V. Voevodsky, V. Goldanskii, N. Emanuel, A. Shilov, A. Merzhanov, and many others. Many of Semenov's students later went on to establish their own scientific schools.

Not only did Semenov dedicate much of his time to teaching young scientists, he also tried to give them positions of power and responsibility, sometimes pushing through resistance from the Academy of Sciences officials.

The decision to appoint A. Dremin, L. Eremenko, G Manelis, A. Merzhanov, L. Stesik, and A. Shilov to lead newly created laboratories in the Chernogolovka branch of ICP was not welcomed by the President of the Academy of Sciences. What could this "*Semenov kindergarten*" possibly achieve in the middle of nowhere without routes and phones?

But Semenov, who became a head of the institute at the age of thirty-five, firmly stood for his appointments. Once these young scientists were appointed as heads of laboratories, Semenov gave them his full support. Manelis recalled how Semenov was a frequent visitor to the first labs and how he shared his so-called "secret" to being a successful organizer in science: *"You have to get together and drink more often".* Soon enough, the President of the Academy of Sciences A. N. Nesmeyanov confessed to others: *"Again Nikolai Nikolayevich is ahead of everyone. In his Chernogolovka base, he is successfully working with the youngest heads of labs and the youngest scientists in the Academy".*[25]

---

[x] Семенов, Н. (2010).Молодежь и будущее. Издательство МГУ.

To be able to select young talented students to become scientific leaders, Semenov initiated the creation of a platform on connecting scientific institutes in the Chernogolovka Center with universities. Such "Academia – University" bases were created at the Moscow Institute of Physics and Technology (also called FizTech or MFTI), Moscow State University, Samara Technical University, Yerevan University, Rostov University, Saratov University, and Tashkent University. In the picture below, from the personal archives of A. Rogachev, a head of the ISMAN lab, Semenov is talking with MFTI President O. Belocerkovsky and students of the Dolgoprudny campus during Semenov's eightieth birthday reception.

We met up with Vilen Azatyan, a corresponding member of the Academy of Sciences of Russia, in his cozy apartment located across the street from the Moscow ICP. Azatyan had a lot to share about Semenov, who had requested his help while working on his book *Development of Chain Reaction Theory and Temperature Processes* in 1967. Many of Azatyan's theoretical and experimental works were dedicated to understanding and managing detonations that occurred during combustion and explosion processes. Safety was always a concern when dealing

Figure 16. President of MFTI O. Belocerkovsky (left) and Semenov (right), 1976. This photo has never been published before. Photo credit – A. Rogachev.

with combustion and explosions, and the practical applications of Azatyan's work led to new safety improvements that would prevent the detonation of natural air and gas mixtures in coal mines and the atomic industry.

For his discoveries, Semenov referred to Azatyan as "the tamer of detonations". Since 1970, Azatyan has written over a dozen other scientific works with Semenov. *"Semenov understood my work,"* remarked Azatyan. *"It was easy to work with him despite scientific arguments, which Semenov was not afraid of."* Azatyan emphasized Semenov's diligence at work, that *"it was often needed to rewrite the entire article after his reviews."* Semenov was also very demanding about the quality and preciseness of experiments. Azatyan recalled his mentor saying that *"subjective interpretation does not belong to science."* [26]

Figure 17. N. Semenov and "the tamer of detonations" Dr. Azatyan, 1976. This photo has never been published before. Photo credit – V. Azatyan.

Figure 18. Dr. Azatyan (right) in his Moscow apartment with the author, Michael Zhilaev, 2019.

Many Soviet physicists taught by Semenov went on to establish their own scientific schools to continue Semenov's legacy of exploration and discovery. One example of such a school is ISMAN, the Institute for Structural Macrokinetics of the Russian Academy of Sciences, created by A. Merzhanov, one of those "legendary four" first heads of labs that started the Chernogolovka Branch of the ICP. In 1967, the Merzhanov laboratory discovered a new group of combustion processes: SHS (Self Propagating High Temperature Synthesis). Created in 1987 from the ICP laboratory of Macrokinetics and the Dynamics of Gas, ISMAN now has two learning centers and many contracts with domestic and foreign universities.[27]

CURRENT RESEARCH AT CHERNOGOLOVKA

Since 2001 Chernogolovka has been recognized as a town, and the population, as of 2019, is about 20,000.

In 2016, the Russian Academy of Sciences academic Vladimir Fortov and his colleagues conducted a fascinating experiment. They attempted to create an explosion modeling the reaction in the center of giant gas planets such as Jupiter. As Fortov explained: "There is always a race to get the highest pressure and temperature to create technical devices. Defense is also important; we must know the effect of the impact of the explosion." According to Fortov,

Figure 19. *Chernogolovka Scientifics Center – Aerial view.*

similar experiments are required to determine how the Earth will evolve in the future.[28]

Currently working in the seven Institutes of Chernogolovka are thirty-eight academics and corresponding members of the Russian Academy of Science, more than three hundred Doctors of Science, and seven hundred PhD researchers. Nobel Prize recipients such as Semenov, Landau, Kapitsa, Alferov, Ginsburg, Abrikosov, and most recently, Andre Geim and Konstantin Novoselov all have close connections with the research center.

Here are just some of the Chernogolovka's recent achievements:

- Development of quantum computers – Institute of Solid State Physics
- New anticancer and organ-regenerating drugs – Institute of Problems of Chemical Physics
- Computational center connected to resources of the most powerful supercomputers of the Russian Academy of Science
- Synthesis and modification of minerals to study the Earth's structure – Institute of Experimental Mineralogy

- Many materials developed in the Institute of Problems of Chemical Physics (formerly the Institute of Chemical Physics) are used everywhere from the military to the space industry.[29]

At the Institute of Structural Macrokinetics Russian Academy of Sciences (ISMAN), scientists are using self-propagating high-temperature synthesis (SHS) to create new materials with unique properties, such as superconductivity or resistance to extreme temperatures. The SHS process was developed by the founder of ISMAN, one of Semenov's pupils, Alexander Merzhanov, by applying combustion processes developed by Semenov. A significant advantage of this process is that materials created using this method can be created in seconds and minutes, as opposed to days using traditional metallurgy.[30]

The "brain drain" of the 1980s and 90s, when many Russian scientists left the country, heavily impacted Chernogolovka and

Figure 20. Demonstration of combustion process at the ISMAN SHS Exhibition. Photo credit – M. Zhilaev.

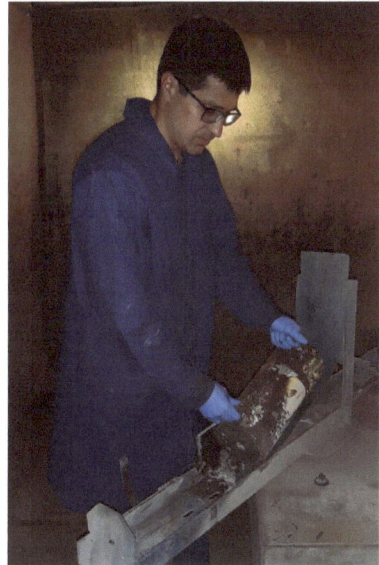

Figure 21. The picture of ISMAN researcher-practitioner, D. Andreev, was taken by the author shortly after conducting a combustion experiment.

science in Russia. Between 1989 and 2004, approximately twenty-five thousand Russian scientists permanently left the country, while thirty thousand worked abroad on temporary contracts. Unsurprisingly, over the last twenty years, the scientific center has shifted to being better known for mass production of vodka and "Chernogolovka" brand soft drinks instead of scientific discoveries. In the scientific institutes, older scientists, many of whom personally knew Semenov and Dubovitsky, now work with young researchers, while most of the middle-aged scientists settled abroad.[31]

ISMAN is one of the institutes trying new approaches to attract young people to work on scientific research. It provides students with modern apartments located just across the road from the institute, as well as many opportunities for promotion, education, and travel. Because of these benefits, many young scientists are coming to ISMAN to start their careers.

CHERNOGOLOVKA IS *Us*!

Sixty years after the founding of Chernogolovka the iconic three-trunked pine at the entrance of the ICP is no longer there, it did not survive the 2016 summer storm. However, the tree is memorialized forever on the city flag, coat of arms, and in monument to the founders in front of the House of Scientists. The children of Chernogolovka, many of whom will go into a scientific career, are now enjoying this monument. They frequently gather here to play, climb, and roller blade.

Semenov's little cottage is located in the woods just 100 meters from the entrance to the ICP. It is now a museum, which preserves the building as it was when Semenov lived there from 1973 to 1986, with many personal documents and items on display, attesting to Semenov's humble persona and his lifelong passion and devotion to science.

Figure 22. *Fragment of Semenov-Dubovitsky Monument. Photo credit – M. Zhilaev.*

Figure 23. *Local kids love playing on Semenov-Dubovitsky monument. Photo credit – M. Zhilaev.*

В ЭТОМ ДОМЕ С 1973 ПО 1986 ГОД
ЖИЛ И РАБОТАЛ ОСНОВАТЕЛЬ
И ПЕРВЫЙ ПРЕДСЕДАТЕЛЬ СОВЕТА ДИРЕКТОРОВ
НОГИНСКОГО НАУЧНОГО ЦЕНТРА РАН В ЧЕРНОГОЛОВКЕ,
ЛАУРЕАТ НОБЕЛЕВСКОЙ ПРЕМИИ,
ДВАЖДЫ ГЕРОЙ СОЦИАЛИСТИЧЕСКОГО ТРУДА,
КАВАЛЕР 9-ТИ ОРДЕНОВ ЛЕНИНА,

АКАДЕМИК
НИКОЛАЙ НИКОЛАЕВИЧ
СЕМЁНОВ

Figure 24. *The memorial sign on the Semenov Museum in Chernogolovka. Photo credit – M. Zhilaev.*

*Figure 25. Semenov office in his cottage house in Chernogolovka
(now the Semenov House Museum). Photo credit – M. Zhilaev.*

*Figure 26. Semenov was using this
typewriter to write the results of his
work in his cottage in Chernogolovka
(now the Semenov House Museum).
Photo credit – M. Zhilaev.*

*Figure 27. Semenov in his cottage –
Chernogolovka, 1970s.
Photo Credit – Semenov Museum.*

Figure 28. Semenov Noble Prize Diploma – the Semenov Museum in Chernogolovka. Photo credit – M. Zhilaev.

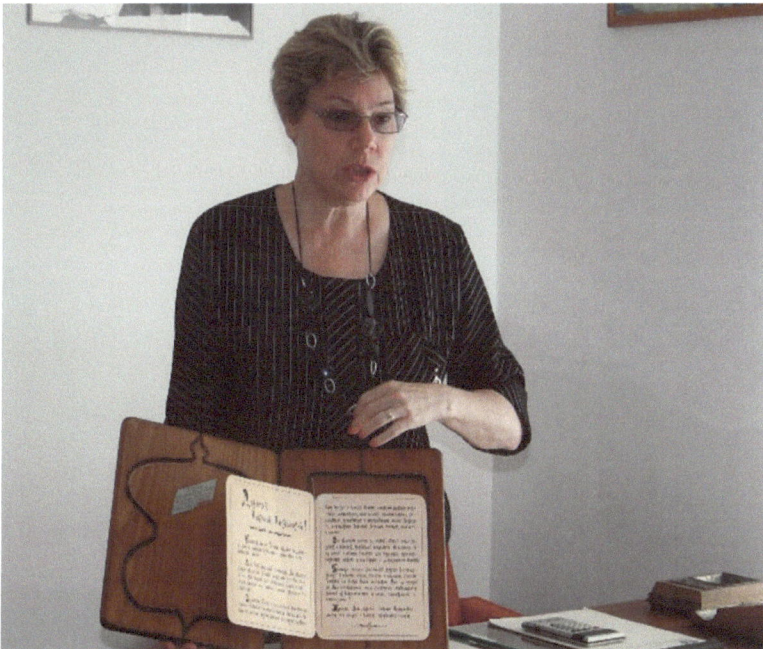

Figure 29. At the Semenov Museum in Chernogolovka. Ms. Tolpygo is presenting Semenov memorabilia. Photo credit – M. Zhilaev.

Figure 30. Building #9 on First Street (Pervaya Ulitsa). Originally built as a hotel, the building became a dorm in the 1980s, during a housing shortage. Photo credit – M. Zhilaev.

Figure 31. One of the first buildings in Noginsk Scientific Center – dormitory "Armyanka" named after students from Erevan University.

ABOUT THE AUTHOR

Michael Zhilaev is a West Windsor-Plainsboro high school student in New Jersey and is interested in history and science. His parents came from Chernogolovka twenty years ago. Both of Michael's parents met Semenov when they were very young researchers at the Institute of Chemical Physics. Pictured below, Michael stands in front of yet another building honoring Semenov at Chernogolovka: a café named after him, located at the Academic Semenov Prospect. This building, containing the Institute library, a large meeting hall, and a cafeteria, was built in 1967, one of the first in Chernogolovka.

Figure 32. Michael Zhilaev is standing in front of one of the oldest building of the Institute of Chemical Physics in Chernogolovka. Now Café Semenov, which is located in this building, displays many scientific devices and other memorabilia of Semenov period in Chernogolovka.

ACKNOWLEDGMENTS

The story of this book began when Princeton University Professor M. Gordin suggested this topic for an independent research project. I am very thankful to him for helping me get started and his advice and encouragement.

While I was writing this story about the outstanding scientist, Nikolai Semenov, I studied many historical materials and interviewed people who personally knew him, including family members. I would like to acknowledge the following scientists from the Institute of Structural Macrokinetics Russian Academy of Sciences (ISMAN) for their willingness to share their experiences, knowledge, and historical and scientific materials: Director Dr. Sanin, head of the Laboratory of Plastic Deformation Materials, Prof. Stolin, head of the Laboratory of Dynamics of Microheterogeneous Processes, Prof. Rogachev, and the head of the ISMAN reading room, Ms. Nasonova.

Especially appreciated is Dr. Azatyan, head of the ICP laboratory, for his thorough insight about Semenov and his help in finding primary sources.

I'm grateful to the Director of the Semenov Museum in Chernogolovka, Ms. Tolpygo, for her enthusiasm and for sharing many pictures and other factual materials.

Many thanks to Semenov's family members for helping me understand that period of life: his daughter Lyudmila Semenova and grandson Alexey Semenov. I would like to acknowledge the special role of Dr. Alexey Semenov who carefully reviewed and verified the facts of this book and shared his constructive feedback to help confirm its accuracy.

Many local materials about Semenov's life in the Volga region were provided by local resident Mr. I. Krivchenkov.

This book would not exist without my family.

Thank you mom, Nadya Zhilaev, for pushing me out of my comfort zone and sending me to Russia to meet with the scientists at Chernogolovka and immerse myself in the scientific environment.

My gratitude to my loving dad, Alexander Zhilaev, for his motivation with high expectations throughout the project.

Thanks to my brother-in-law, Harvey Chin, and my niece, Natalie Chin, for their love and encouragement.

My enormous gratitude to Olga Chin, my older sister, who gave me the inspiration for this project. In the spring of 2019, she connected me with Princeton University Professor Gordin, which began my journey into research and writing. With helpful advice and words of motivation, she made sure I could do everything the best I could.

I appreciate my teachers at West Windsor-Plainsboro High School South, who gave me a love of learning and supported me in everything.

Learning the stories of Soviet scientists taught me the meaning of patience, integrity, and dedication to knowledge.

I'm eternally grateful to these scientists for their enormous sacrifices.

APPENDIX

APPENDIX1.

Protocol of the Special Nuclear Committee meeting signed by Beria to execute the plan for first atomic bomb nuclear test on Aug 29, 1949.

APPENDIX 2.

The Soviet Decree on starting the construction of the ICP testing grounds.

Приложение № 2

<u>Для служебного пользовани:</u>

СОВЕТ МИНИСТРОВ СССР

РАСПОРЯЖЕНИЕ

от 28 февраля 1956 г. № 1024-рс

Москва, Кремль

<u>ВЫПИСКА</u>

2. Обязать Министерство обороны СССР (т.Жигарева) выделить на территории действующего полигона Военно-Воздушной инженерной академии имени Жуковского (Ногинский район, Московской области) земельный участок площадью 20-25 кв.км для создания Научно-исследовательского полигона при Институте химической физики Академии наук СССР.

3. Разрешить Академии наук СССР вести строительство лабораторных, опытно-производственных и жилых об'ектов Научно-исследовательского полигона при Институте химической физики (Ногинский район, Московской области) с общим об'емом капиталовложений в 1956-1958гг. в сумме 25,6 млн.рублей.

Зам. Председателя
Совета Министров Союза ССР   М.Первухин

Верно
Начальник отдела по обеспечению
деятельности Архива Президента
Российской Федерации   В.Якушев

26.04.2002

Appendix 3.

The Presidium of Soviet Academy of Sciences Decree on organizing and building the Noginsky Science Center in Chernogolovka dated August 9, 1962.

ПРЕЗИДИУМ АКАДЕМИИ НАУК СОЮЗА ССР

РАСПОРЯЖЕНИЕ   *10390*

№ 141-1275 от 9 августа 1962 г.

О мероприятиях по строительству и организации работ комплекса институтов в Черноголовке (Ногинский научный центр АН СССР).

1. Для координации научных и научно-организационных работ по Ногинскому научному центру АН СССР учредить Совет директоров в составе:

академик Н.Н.Семенов - председатель

академик А.П.Виноградов

академик Н.М.Жаворонков

академик В.А.Кириллин

академик Г.В.Курдюмов

академик Д.С.Коржинский

член-корр.АН СССР Б.К.Вайнштейн

доктор хим.наук Ф.И.Дубовицкий

ученые секретари Совета, к.х.н.Л.Г.Щербакова,Г.С.Петелина.

2. Создать при Совете директоров организационную комиссию для координации работ по строительству и административно-хозяйственному управлению Ногинского научного центра АН СССР в составе:

д-р хим.наук Ф.И.Дубовицкий - председатель

             Б.П.Золотой   - зам.председателя

кан.физ.мат.наук Ю.Я.Осипян

д-р тех.наук А.Е.Шейндлин

             М.А.Щусев   - ГИПРОНИИ АН СССР

             Н.Н.Кулигин   - Центральное управление капитального строительства АН СССР

зам.директора института Новых химических проблем, Кристаллографии и ГЕОХИ.

- 2 -

3. Поручить Филиалу Института химической физики АН СССР осуществлять функции генерального заказчика по строительству Ногинского научного центра АН СССР.

4. Назначить т.Дубовицкого Ф.И. уполномоченным Президиума АН СССР по организационно-хозяйственным вопросам Ногинского научного центра и т.Золотого Б.П. - его заместителем.

5. Для временного размещения Института новых химических проблем и Института физики твердого тела:а) предложить Филиалу Института химической физики (т.Дубовицкий Ф.И.) предоставить этим институтам впредь до осуществления строительства их корпусов

лабораторный корпус на 2-й промплощадке (2200 кв.м. полезной площади по окончанию строительства) для лабораторных работ;

700 кв.м. в полимерном корпусе для укрупненных установок этих институтов;

б) т.Чернопятову К.Н. и Дубовицкому Ф.И. обеспечить окончание строительства лабораторного корпуса на 2-й промплощадке к I сентября 1963 г.,т.Долгополову В.Н. обеспечить в 1963 г.указанные корпуса соответствующим оборудованием за счет смет Института новых химических проблем и Института физики твердого тела на 1963 год;

в) взамен помещений, предназначавшихся в лабораторном корпусе на 2-й промплощадке для размещения бесфоновой лаборатории (I200 кв.м.),построить в течение 1962-63 г.г. для ГЕОХИ отдельно стоящую бесфоновую лабораторию по ранее утвержденному проектному заданию.

6.Академикам Н.М.Жаворонкову и Г.В.Курдюмову в 2-недельный срок:

а) дать предложения о перспективном плане развития научных направлений в Институте новых химических проблем и Институте физики твердого тела на 1963-65 г.г.

б) дать предложения о потребности в научных кадрах по годам до 1965 г. включительно и мероприятиях по подготовке и отбору кадров

- 3 -

в) дать заявки на оборудование, необходимое для оснащения временно передаваемых им в 1963 г. филиалом Института химической физики АН СССР помещений;

г) уточнить и передать в ЦУКС плановые задания на проектирование институтов, а также соображения о применении типовых или повторных корпусов и вводе отдельных объектов по годам (1963-65 г.г.).

Пункты 6а,б,в,г рассмотреть и утвердить на Совете директоров Ногинского научного центра.

7. Поручить ГИПРОНИИ (Доморацкий П.И.):

а) до 1 января 1963 г. разработать генеральный план Ногинского научного центра;

б) в 1 квартале 1963 г. разработать комплексное проектное задание по Институту физики твердого тела;

в) в течение 1 полугодия 1963 г. разработать комплексное проектное задание Института новых химических проблем;

г) обеспечить до конца 1962 г. проектную документацию, необходимую для строительства 1963 г. объектов филиала ИХФ, а так же подготовительные работы по строительству новых институтов;

д) закончить создание и укомплектование мастерской ГИПРОНИИ в Черноголовке к 1 октября 1962 г.

8. Поручить уполномоченному Президиума АН СССР по строительству т.Чернопятову К.Н.:

а) совместно с ЦУКСом и с привлечением академиков Семенова Н Жаворонкова Н.М., Курдюмова Г.В. и т.Золотого Б.П. подготовить соображения о стоимости строительства Ногинского научного центра и распределения финансирования по годам;

б) учитывая, что строительство указанных институтов должно быть начато уже в 1963 г., оформить разрешение на производство и оплату выполняемых строительно-монтажных работ по локальным сметно-финансовым расчетам, впредь до утверждения комплексных проектных заданий.

- 4 -

9. Поручить Институту химической физики согласовать с
Главспецстроем объем строительно-монтажных работ на 1963 г.
на сумму 7 млн.руб., в том числе 2 млн.руб. на строительство
новых институтов, перевалочного лабораторного корпуса и бесфоновой лаборатории.

10. Разрешить филиалу Института химической физики АН СССР
временно производить оплату проектно-сметных работ по научному
центру за счет средств, предусмотренных на проектные работы в
сводной смете на строительство филиала ИХФ, с последующим вос-
становлением всех затрат за счет сметы на строительство новых
институтов.

11. Для улучшения качества отделочных и спецмонтажных
работ разрешить филиалу Института химической физики построить
в 1963 г. один 64-квартирный жилой дом за счет сводной сметы на
строительство филиала, заселив его высококвалифицированным воль-
нонаемным составом строителей Главспецстроя.

И.о.Президент
Академии наук СССР
академик   -М.В.Келдыш

AH CCCP-1649-4, т.50
10.УШ.62г. № 13

APPENDIX 4.

The Decree of the Presidium of Soviet Academy of Sciences on organizing and constructing additional scientific institutes: The Institute of Solid State Physics and the Institute of New Chemical Problems.

АРХИВНАЯ КОПИЯ.

247

ПРЕЗИДИУМ АКАДЕМИИ НАУК СОЮЗА ССР

РАСПОРЯЖЕНИЕ №40-1273

«8» августа 62
г. Москва 195 г.

Совет Министров СССР распоряжением № 1855р от 7 июля 1962 года:

1. Принял предложение Академии наук СССР и Государственного комитета Совета Министров СССР по координации научно-исследовательских работ, согласованное с Госпланом СССР и Госэкономсоветом СССР:

а/ о строительстве, в виде исключения, в Московской области научно-исследовательских институтов:

— Института физики твердого тела /Ногинский район, Московской области/ в 1963-1965 г.г.

— Института новых химических проблем /Ногинский район, Московской области/ в 1964-1967 г.г.

— Института физики высоких давлений /район Красной Пахры, Московской области/ в 1963-1965 г.г.

б/ об установлении сроков строительства научно-исследовательских учреждений научного городка в с.Пущино, Серпуховского района, Московской области:

—Института белка в 1963-1964 г.г.

—Института фотосинтеза в 1964-1965 г.г.

—Института биохимии и физиологии микроорганизмов в 1964-1965 г.г.

—СКБ биологического приборостроения в 1963-1964 г.г.

Разрешил Академии наук СССР осуществлять в 1962-1963 г.г. строительство научного городка в с.Пущино по проектам и сметам на отдельные объекты.

2. Обязал Академию наук СССР утвердить до 1 января 1964 года генеральную смету на строительство биологических институтов в с.Пущино в установленном порядке.

2-я тип. Издат. АН СССР. Москва. Зак. 7750 Тир. 10 000

– 2 –

Предложить

3. Госплану СССР и Государственному комитету Совета Министров СССР по координации научно-исследовательских работ предусматривать в проектах годовых народнохозяйственных планов выделение капитальных вложений и ассигнований на строительство научно-исследовательских учреждений, предусмотренных распоряжением.

В соответствии с распоряжением Совета Министров СССР от 7/УП-1962 г. за № 1855р обязать:

I. Отделения АН СССР представить в ЦУКС задания на проектирование:

а/ Отделение химических наук (академик Н.Н.Семенов) Института новых химических проблем к 20 августа 1962 года,

б/ Отделение физико-математических наук (академик Л.А.Арцимович) Института физики твердого тела к 20 августа 1962 года,

в/ Отделение биологических наук (академик Н.М.Сисакян) Института фотосинтеза, Института физиологии и биохимии микроорганизмов к I октября 1962 года.

2. ЦУКС АН СССР (Г.И.Русановский):

а/ Предусматривать, начиная с 1963 года, в планах капитальных вложений АН СССР ассигнования на строительство научно-исследовательских учреждений, предусмотренных настоящим распоряжением,

б/ включить в планы проектных работ ГИПРОНИИ на 1962 и 1963 годы проектирование этих научно- исследовательских учреждений,

в/ представить на утверждение Президиума АН СССР генеральную смету на строительство биологических институтов в с.Пущино к I декабря 1963 года,

г/ совместно с Центракадемстроем (т.Совков Г.В.), ГИПРОНИИ (т.Доморацкий П.И.), институтами-заказчиками и Уполномоченным ЦУКСа по Москве и Московской области, разра-

- 3 -   249

ботать и представить в месячный срок уполномоченному президиума АН СССР по строительству К.Н.Чернопятову, мероприятия
по выполнению настоящего распоряжения.

3. Институт химической физики АН СССР (академик
Н.Н.Семенов) выполнять обязанности заказчика-титулодержателя
по строительству Института физики твердого тела и Института
новых химических проблем, а также по жилищному и культурно-
бытовому строительству для этих Институтов.

4. Институт физики высоких давлений (член-корреспондент
АН СССР Л.Ф.Верещагин) выполнять обязанности заказчика-
титулодержателя по строительству Института физики высоких
давлений и лаборатории спектроскопии, а также жилищному и
культурно-бытовому строительству в Красной Пахре.

5. Контроль за выполнением настоящего распоряжения
возложить на уполномоченного Президиума АН СССР по строительству тов. К.Н.Чернопятова.

Президент Академии наук СССР
академик   М.В.КЕЛДЫШ

и.о.главного ученого секретаря
президиума Академии наук СССР
член-корреспондент АН СССР   М.И.АГОШКОВ

6.7.62

ОСНОВАНИЕ: Архив РАН. Ф.2, оп.13, д.315, л.л.247-
-249.

Директор Архива   (Б.В.Левшин)

Научный сотрудник   (Н.К.Ткачен

Г.Русановский
24/VII-62 г.

INDEX

BIBLIOGRAPHY

1 Аникин, В., Усанов. Д. (2016). Николай Николаевич Семенов: Волжские сюжеты жизни. Известия Саратовского Университета. Саратов.

2 The Lord Dainton. (September 1986). Article Nikolay Nikolayevich Semenov. Biographical Memoirs of Fellows of the Royal Society. Vol. 36.

3 Аникин, В. Хроника из истории физики. К 120-летию со дня рождения академика Н. Н. Семёнова физик-инноватор, земляк, учитель и друг Н. Н. Семёнова Владимир Иванович Кармилов. Известия Саратовского Университета. Новая Физика. 2016, 1, p 44-54.

4 Hargittai, I. (2013). Buried Glory: The Portrait of Soviet Scientists. Oxford University Press.

5 Семенов, Н. (2010).Молодежь и будущее. Издательство МГУ.

6 Под Редакцией Академика Андреева, А. (1998). Капица. Тамм. Семенов. Вагриус. Москва.

7 Гольданский, В. (1993). Фрагменты Минувшего. Сборник воспоминаний об Академике Николае Николаевиче Семенове. Москва, Р. 62.

8 Semenova, L. (2016). Diary of Semenov daughter, Online magazine "Семь Искусств".

9 Харитон, Ю. (1993). НАЧАЛО. Сборник воспоминаний об Академике Николае Николаевиче Семенове. Р. 31-42.

10 Pace, E., (Dec 20 1996). Khariton, Father of Soviet Atom Bomb, Dies at 92. New York Times.

11 Khariton, Y., Walta, Z. (1926). Oxydation von Phosphordämpfen bei niedrigen Drucken. – Z. Phys., Bd 39, H. 7– 8, S. 547–556.

12 Semenov, N., Frenkel Y. (1928). On the theory of combustion processes.1. J.Rus. Phys. Chem. Soc. Ser. Phys., Vol.60, No.3, P.241-250.

13 Semenov, N. (1930). Kinetics of chain reactions. Trans. Leningrad Phys. Techn. Lab., No.14, P.13-25.

14 Semenov, N. (1940). Thermal theory of combustion and explosion: Introduction. Part 1, 2. Sov. J. Advances in Phys. Sci., Vol.23, No.3, P.251-292.

15 Semenov, N. (1940). Thermal theory of combustion and explosion. Part 3. Sov. J. Advances in Phys. Sci., Vol.24, No.4, P.433-486.

16 Адушкин, В., Сулимов, А.(2019). Вклад ученых Химфизики в советский Атомный проект. Торус пресс, Москва.

17 Харитон, Ю.(8 декабря 1992). Ядерное оружие СССР: пришло из Америки или создано самостоятельно? Газета «Известия».

18 Харитон, Ю., Смирнов Ю. (1999) О некоторых мифах и легендах вокруг создания Советского атомного и водородного проектов. В сб. «Человек столетия Ю. Б. Харитон». Москва.

19 Садовский, М. (1997). Институт Химической Физики. В Сб. "Курчатовский институт. История атомного проекта." Вып. 11. С.45-64.

20 Чернышев, А. (2016). Творец истории XX века Николай Николаевич Семенов в атомном проекте СССР. Торус Пресс. Москва.

21 Леенсон, И. (2016) Физик, ставший химиком: Николай Николаевич Семёнов, «Троицкий вариант» №214-215.

22 Адушкин, В. (2016). Институт химической физики — колыбель ядерного оружия России // Наука и технологии в промышленности. № 1. P. 31–36.

23 Дубовицкий, Ф. (1992). Институт химической физики: очерки истории. Черноголовка.

24 Rogacheva, M. (2017). The Private World of Soviet Scientists from Stalin to Gorbachev. Cambridge University Press.

25 Манелис. Г . (2011). Химфизики. – Черноголовка: Редакционно-издательский отдел ИПХФ РАН.

26 Азатян, В. (1993). В Лаборатории Семенова. Сборник воспоминаний об Академике Николае Николаевиче Семенове. P. 79-81.

27 Столин, А., Юхвид В., Алымов, М. (2017). Академический институт и его история. Черноголовская Газета. #15-16.

28 Лескова, Н. (21 декабря 2016). В Черноголовке зажгли сверхновую. Портал Научная Россия.

29 Столина, А. (20 июня 2018 г.). О современных разработках и технологических достижениях Научного центра РАН в Черноголовке. Черноловская Газета.

30 Мержанов, А. (2005). Лучше быть нужным, чем свободным. Рос. акад. наук, Науч. совет по горению и взрыву. – Черноголовка.

31 Киреев, М. (08.10.2010). Исход Ученых Умов. Der Spiegel.

www.ingramcontent.com/pod-product-compliance
Lightning Source LLC
Chambersburg PA
CBHW041721200326

41521CB00004B/166